What are solid shapes?
Come along and see!
Math that has to do with shapes
is called geometry.

These are solid shapes.
Each one takes up space.
Do you see the sides they have?
Each one is called a face.

Sphere

Rectangular Prism

Pyramid

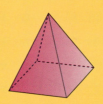
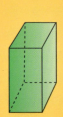

Try to find the faces made of flat shapes that you know— triangle, rectangle, circle, square. Ready, set, let's go!

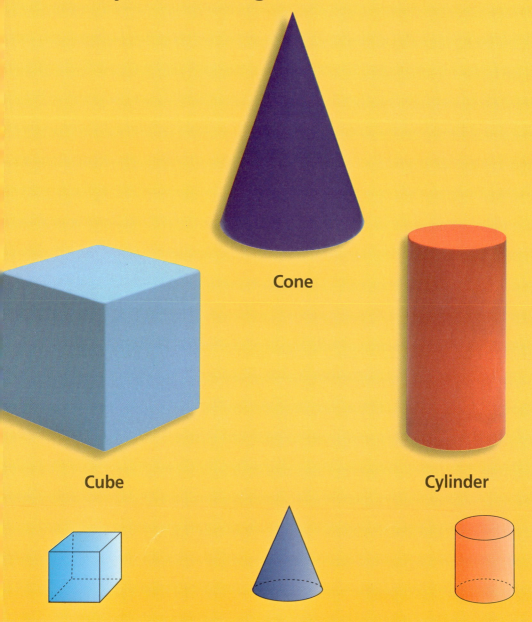

Cone

Cube

Cylinder

Solid shapes are everywhere. Look at the solid shapes here.

Pyramid

Cones

Cube

Pyramid, cone, cube, cylinder, rectangular prism, and sphere!

Cylinder

Rectangular Prism

Sphere

What solid shapes are towering here?

Name all the shapes you see.

What solid shape is this soccer ball?

What could this icy shape be?

Here are two solid shapes that you should know—
the cones hold a treat
and the cylinders grow!

What are the names of these solid shapes and the flat shapes on each side?

It's easy to spot the triangles, but where are the squares? Can you decide?

Look at all these solid shapes. Can you name each one?

Learning the names of all the shapes can make geometry fun!